Now I Know

# Clouds

Written by Roy Wandelmaier
Illustrated by John Jones

**Troll Associates**

*Library of Congress Cataloging in Publication Data*

Wandelmaier, Roy.
  Clouds.

  Summary: Brief text and illustrations explain the
characteristics of different kinds of clouds.
  1. Clouds—Juvenile literature.  [1. Clouds]
I. Jones, John, 1935-    ill.  II. Title.
QC921.5.W36    1985      551.57'6       84-8643
ISBN 0-8167-0338-8 (lib. bdg.)

Look up at the sky.

What do you see?

Clouds.

What is a cloud?

Tiny drops of water, or ice, floating in the air.

Clouds help us guess what the weather will be.

What kinds of clouds are there?

*Stratus* clouds float low in the sky, flat as sheets.

They may bring rain or drizzle.

*Cumulus* clouds are white,
and pile high in the air.

That means fair weather is coming.

*Cirrus* clouds are white and curly.

They float highest of all,
and bring a change of weather.

When any of these clouds turn dark,
we call them *nimbus*. That means rain!

Or snow, if it is cold.

When a cloud lies on the ground,
we call it fog.

Just be careful when you walk through it!

You can also visit clouds by flying through them.

Or above them.

Don't get caught in a cloudburst!

Who needs all this rain anyway?

Plants do.
And so do farmers.

Others just enjoy it.

Isn't it fun to watch clouds?

What do you see there?

A horse? A dragon?

Maybe even a face.

Winds blow clouds many miles across the sky.

When do clouds look best?

When you look up and
you see something wonderful.